Asthma: Clear Answers and Smart Advice for Someone Diagnosed with Asthma

By Stacey Chillemi

Published by Stacey Chillemi at Lulu

I0475137

Lulu Edition, License Notes

Contents

Preface

Managing the Stress of Asthma can be devastating for the people and their families who are affected with the lifelong disease. Many people who have been diagnosed with asthma feel distressed frustrated, and very scared. Today, approximately 300 million people worldwide suffer from asthma, with 250,000 annual deaths attributed to the disease. It is estimated that the number of people with asthma will grow by more than 100 million by 2025. It is one of this world's most common and costly diseases. There is no cure for asthma, but asthma can be managed with proper prevention and treatment.

Inside this book, you'll discover...

- Valuable information about asthma
- How to treat & care for your asthma
- The support you need
- Self-help techniques to help you cope emotionally
- Recovery techniques
- How your diet affects your asthma condition and what you can do to improve it?
- The holistic approach to treating asthma
- How to cope with your asthma emotionally
- Worldwide support resources for asthma
- And much more…

This guide will teach you how you can improve your asthma condition, so you can be healthy and avoid any unnecessary asthma attacks. This book will supply you with all the necessary information to understanding asthma and the step-by-step techniques on how to manage, treat and cope with the disease.

Introduction

This book is an overview of asthma a disease that afflicts millions of people worldwide.

Topics include - What asthma is, how doctors diagnosed asthma, how the various symptoms, such as shortness of breath, wheezing and tightness in the chest, manifest themselves in different individuals.

A large portion of the book is devoted to a lengthy and highly useful discussion of what it is, the current therapies available, coping strategies, asthma resources, as well as the pros and cons of specific medications. Other topics in this book include alternative and complementary therapies, the early warning signs of asthma and asthma attacks.

Most important the book goes into detail about triggers and how to avoid them. The book also discusses other important topics such as, asthma caused by allergens and environmental pollutants. These topics are identified, along with suggestions for avoiding or alleviating them.

Section One – Learning the Basics

Chapter 1: The Facts About Asthma

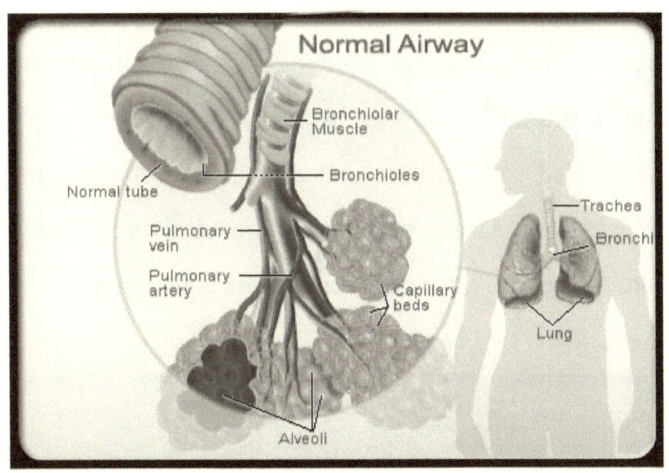

Asthma is a chronic disease of the airways that makes breathing difficult. With asthma, there is inflammation of the air passages that result in a temporary narrowing of the airways that carry oxygen to the lungs. This results in asthma symptoms, including coughing, wheezing, shortness of breath, and chest tightness. Some people refer to asthma as *"bronchial asthma."*

Even though there are seemingly miraculous treatments for asthma symptoms, asthma is still a serious -- even dangerous -- disease that affects more than 22 million Americans and causes nearly two million emergency room visits ever year. With proper asthma treatment, you can live well with this health condition. Yet inadequate asthma treatment limits the ability to exercise and be active. Poorly controlled asthma can lead to multiple visits to the emergency room and even hospital admission, which can affect your performance at home or at work.

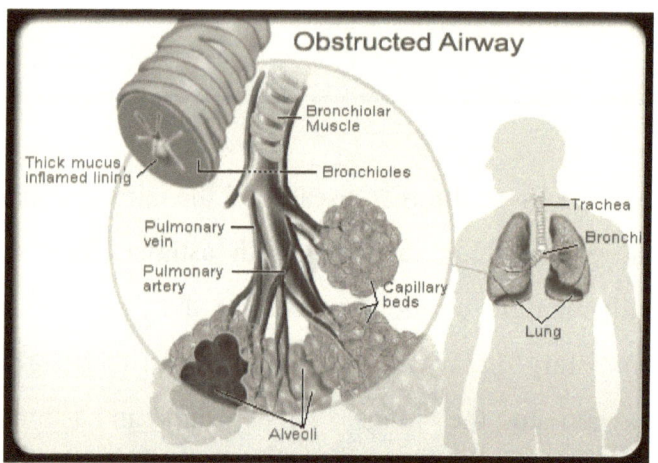

There are three main features of asthma:

1. **Airway obstruction:** During normal breathing, the bands of muscle that surround the airways are relaxed, and air moves freely. However, in people with asthma, allergy-causing substances and environmental triggers make the bands of

muscle surrounding the airways tighten, and air cannot move freely. Less air causes a person to feel short of breath, and the air moving through the tightened airways causes a whistling sound known as wheezing.(Fortunately, this airway narrowing is reversible, a feature that distinguishes asthma from other lung diseases such as bronchitis or emphysema.)

2. **Inflammation:** People with asthma have red and swollen bronchial tubes. This inflammation is thought to contribute greatly to the long-term damage that asthma can cause to the lungs. Therefore, treating this inflammation is key to managing asthma in the end.

3. **Airway irritability**: The airways of people with asthma are extremely sensitive. The airways tend to overreact and narrow due to even the slightest triggers such as pollen, animal dander, dust or fumes.

Chapter 2: What Can Cause Asthma?

It is not clear exactly what makes the airways of people with asthma inflamed. Your inflamed airways may be due to a combination of things. We know that if other people in your family have asthma, you are more likely to develop it. New research suggests that being exposed to things like tobacco smoke, infections, and some allergens early in your life may increase your chances of developing asthma.

There are things in the environment that cause asthma symptoms and lead to asthma attacks.

Below is a list that gives some examples of things that can bring on asthma.

- Allergens
- Animal dander (from the skin, hair, or feathers of animals)
- Dust mites (contained in house dust)
- Cockroaches
- Pollen from trees and grass

- Mold (indoor and outdoor)
- Irritants
- Cigarette smoke
- Air pollution
- Cold air or changes in weather
- Strong odors from painting or cooking
- Scented products
- Strong emotional expression
- Crying
- Laughing hard
- Stress
- Medicines
- Aspirin
- Sulfites in food
- Dried fruit
- Beverages
- Wine
- Exercise
- Allergens
- Irritants
- Viral infections
- Exercise

- Some people only have asthma after a viral infection.

A condition called gastro esophageal, a reflux disease that causes heartburn and can worsen asthma symptoms, especially at night. Irritants or allergens that you may be exposed to at your work, such as special chemicals or dusts Infections

This is not a complete list of all the things that cause asthma in a child or an adult. The list touches confers examples the main things that can bring on asthma. People can have trouble with one or more of these. It is important for you to learn which ones are problems for you. Your doctor can help you identify which things affect your asthma and ways to avoid them.

Chapter 3: Can You Get Asthma As An Adult?

Asthma affects 22 million Americans. Asthma in children occurs in approximately 10%-12% of kids. Asthma may occur at any age, although it is more common in younger individuals under the age of forty.

People who have a family history of asthma have an increased risk of developing the disease. Allergies and asthma often occur together. Smoking with asthma is a dangerous combination. Surprisingly, many older individuals who have asthma smoke.

People with allergies, such as hay fever or animal allergies, often get asthma. For example, if you are allergic to dogs, you might have an asthma attack if you get near a dog. If something at your workplace gives you an asthma attack, it is called work-related asthma.

Some of the work conditions that can cause asthma attacks are cold temperatures, heavy physical work, dust, chemicals, and smoke. Some people first get allergies to chemicals or dust in the workplace, and then they get asthma later on.

Anyone can develop asthma at any time, and adult-onset asthma happens frequently. If you have symptoms of asthma, talk to your doctor. If you have adult-onset asthma, your doctor will instruct you in using the asthma inhalers and other asthma medications to prevent further breathing problems.

Chapter 4: What are the Symptoms of Asthma for Adults and Children?

People with asthma experience symptoms when the airways tighten, inflame, or fill with mucus.

Common symptoms of asthma include:

- Coughing, especially at night
- Wheezing
- Shortness of breath
- Tightness in the chest
- Pain in the chest
- Pressure in the chest

Still, not every person with asthma has the same symptoms in the same way. You may not have all of these symptoms, or you may have different symptoms at different times. Your asthma symptoms may also vary from one asthma attack to the next, being mild during one asthma attack and severe during another.

Some people with asthma may go for extended periods without having any symptoms, interrupted by periodic worsening of their symptoms called asthma attacks. Others might have asthma symptoms every day. In addition, some people with asthma may only have asthma during exercise or asthma with viral infections like colds.

Mild asthma attacks are generally more common. Usually, the airways open up within a few minutes to a few hours. Severe attacks are less common but last longer and require immediate medical help. It is important to recognize and treat even mild symptoms to help you prevent severe episodes and keep asthma under better control.

The Early Warning Symptoms of Asthma

Early warning signs are changes that happen just before or at the very beginning of an asthma attack. These asthma attack symptoms may start before the well-known symptoms of asthma and are the earliest signs that your asthma is worsening.

In general, these signs are not severe enough to stop you from going about your daily activities. However, by recognizing these signs, you can stop an asthma attack or prevent one from getting worse.

Early warning signs include:

- Frequent cough, especially at night
- Losing your breath easily or shortness of breath
- Feeling very tired or weak when exercising
- Wheezing or coughing after exercise
- Feeling tired, easily upset, grouchy, or moody
- Decreases or changes in lung function as measured on a peak flow meter

- Signs of a cold, or allergies (sneezing, runny nose, cough, nasal congestion, sore throat, and headache)
- Trouble sleeping

Asthma in Children

Asthma is increasingly prevalent among children. Nearly one in 10 American children now have asthma, a sharp rise that still has scientists searching for a cause. An estimated 6.5 million children under age 18 (8.9%) are now diagnosed with the disease. The rate of childhood asthma has more than doubled since 1980, according to the CDC.

Asthma symptoms can vary from episode to episode in the same child. Signs and symptoms to look for include:

- Frequent coughing spells, which may occur during play, at night, or while laughing or crying
- A chronic cough (which may be the only symptom)
- Less energy during play

- Rapid breathing (intermittently)
- Complaint of chest tightness or chest "hurting"
- Whistling sound when inhaling or exhaling.
- This whistling sound is called wheezing.
- Seesaw motions in the chest from labored breathing. These motions are called retractions.
- Shortness of breath
- Loss of breath
- Tightened neck
- Tighten chest muscles
- Feelings of weakness
- Tiredness

While these are some symptoms of asthma in children, your child's doctor should evaluate any illness that complicates your child's breathing. About half of infants and toddlers with repeated episodes of wheezing with shortness of breath or cough (even though these illnesses usually respond to asthma medications) will not have asthma by the age of six. Because of this, many pediatricians use terms like ***"reactive airways disease"*** or bronchiolitis when describing such conditions on children. They use those terms instead of calling them asthmatic.

Chapter 5: Diagnosing Asthma

Knowing all about your asthma triggers and symptoms can help your doctor make an accurate diagnosis and prescribe the most effective treatment. Learn about the tests that your doctor may use to make an asthma diagnosis. Discover more about lung or pulmonary function tests and the different tests used for allergy and asthma.

Problems with Diagnosing Asthma

The problem with diagnosing asthma is most of the time patients do not have obvious asthma symptoms when they arrive at the doctor's office.

For instance, you may have coughed and wheezed for a week, and by the time, you see your doctor, you have no symptoms at all. Then suddenly, when you least expect it, you might have asthma attack symptoms such as shortness of breath, coughing, and wheezing. Sometimes allergies to seasonal pollen or weather changes can trigger asthma attack symptoms. Other times, a viral infection such as cold or flu can trigger asthma attack symptoms. Smoking with asthma can worsen asthma symptoms, as can sinusitis and asthma. Even exercise or sudden stress or allergies to aspirin or other medications can cause asthma attack symptoms.

If you have asthma, you may go for weeks to months without having any asthma symptoms. That makes diagnosing asthma even more difficult -- unless you do some homework, figure out your asthma triggers and causes of asthma, and help your doctor make an accurate asthma diagnosis. Once an accurate diagnosis is made, you can learn to recognize and treat your asthma symptoms with the right asthma medications so asthma does not interrupt your busy life.

Asthma triggers fall into two categories:

- Allergens ("specific")
- Nonallergens -- mostly irritants (nonspecific)

Your doctor may use one or more of the following asthma tests in diagnosing asthma. These tests are used to assess your breathing and to monitor the effectiveness of asthma treatment.

Spirometry

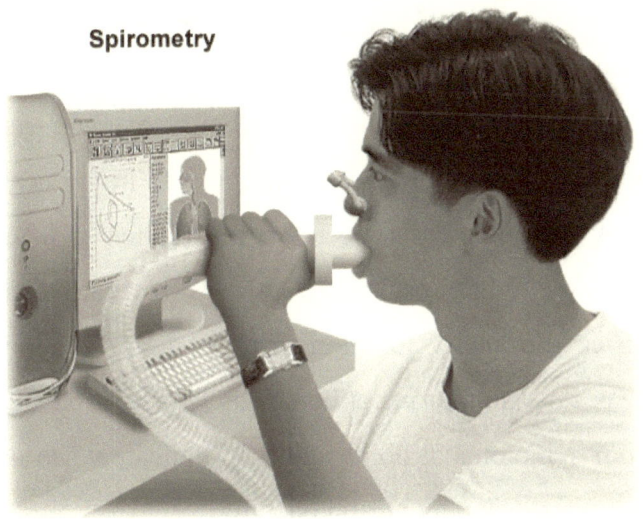

Spirometry -- a pulmonary function test that measures how much air you can exhale. This asthma test confirms the presence of airway obstruction that improves with treatment, which is very characteristic of asthma, and can accurately measure the degree of lung function impairment. This test can also monitor your response to asthma medications and is recommended for adults and children over age five.

Peak Flow Testing – a self-assessment you can do at home to evaluate lung function. *The peak flow rate (PEFR)* provides a reliable objective measure of airway function. Your doctor will go over how to use a peak flow meter, which involves taking a deep breath and blowing out as hard as you can. Peak flow is the highest airflow velocity that you can achieve. When done accurately, a drop in the peak flow measurement reflects an obstruction in your airways. While peak flow is less accurate than office spirometry for monitoring of lung function, peak flow monitoring at home can help you manage your symptoms at home and help indicate when an asthma attack may be approaching.

Chest X-Ray – while not routinely required, if there are symptoms that may be caused by another condition such as pneumonia, your doctor may want to do a chest X-ray. On the other hand, if your asthma treatment is not working as well as it should, a chest X-ray may help to clarify the problem.

In diagnosing asthma, your doctor may order other asthma tests, including a methacholine challenge test. Methacholine is an agent that, when inhaled, causes airways to spasm and narrow if asthma is present.

Not everyone needs every asthma test. Trust your doctor to decide which set of asthma tests is best in your case to ensure no other medical problems are present. This can help you avoid extra testing that may add little to your diagnosis and only increases the number of tests and expense. If you still do not feel comfortable with the asthma diagnosis, talk to your doctor to see if more testing is necessary. On the other hand, get a second opinion until you have peace of mind that the asthma or breathing problem has been diagnosed correctly. Then, proper asthma treatments can begin.

Getting back in control of your asthma depends on an accurate asthma diagnosis and asthma support. Once asthma is properly diagnosed, your doctor can prescribe the most effective asthma treatments, including an asthma inhaler and inhaled steroids that can relieve your breathing problems and help with prevention of asthma symptoms.

Chapter 6: What Is An Asthma Attack?

An asthma attack is the episode in which bands of muscle surrounding the airways are triggered to tighten. This tightening is called bronchospasm. During the attack, the lining of the airways becomes swollen or inflamed and the cells lining the airways produce more and thicker mucus than normal.

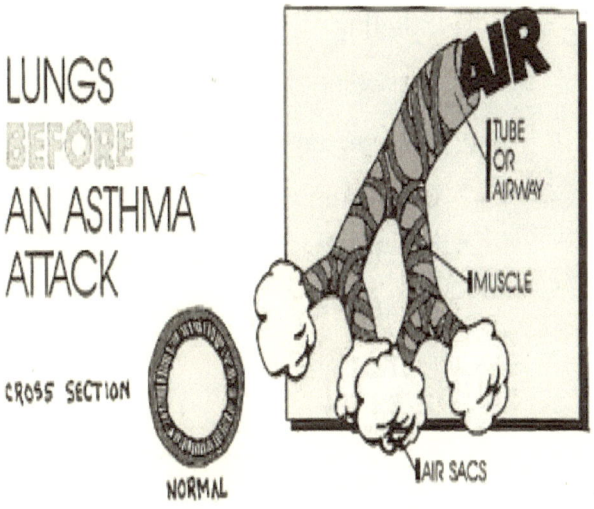

All of these conditions -- **bronchospasm, inflammation**, and **mucus production** -- cause symptoms such as:

- Difficulty in breathing

- Wheezing
- Coughing
- Shortness of breath
- Difficulty performing normal daily activities

Other symptoms of an asthma attack include:

- Severe wheezing when inhaling and exhaling
- Persistent coughing
- Extreme rapid breathing
- Chest pain or pressure
- Tightened neck and chest muscles, this is called retractions
- Difficulty talking
- Feelings of anxiety
- Feelings of panic
- Pale face
- Sweaty face
- Blue lips
- Blue fingernails

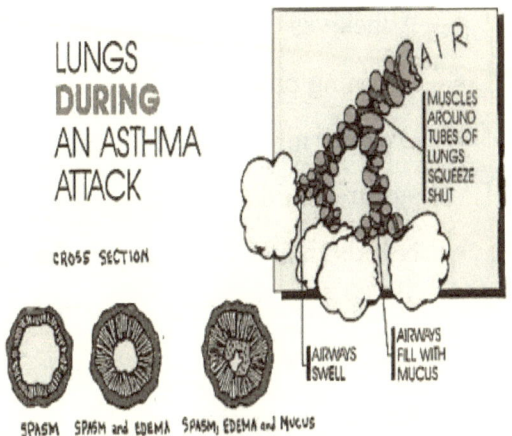

LUNGS **DURING** AN ASTHMA ATTACK

CROSS SECTION

AIR

MUSCLES AROUND TUBES OF LUNGS SQUEEZE SHUT

AIRWAYS SWELL

AIRWAYS FILL WITH MUCUS

SPASM   SPASM and EDEMA   SPASM, EDEMA and MUCUS

Allergens (substances that cause asthma attacks:

- Dust and dust mites

- Pollen

- Grass

- Mold

- Tobacco smoke (or smoke from burning wood, leaves, paper, etc.)

- Animal dander

- Feathers

- Cockroaches and related insects

- Chemicals and chemical fumes

- Including perfumes

- Room deodorizers

- Paint

- Adhesives
- Cleaning products
- Outdoor air pollution (such as ozone and sulfur dioxide)
- Nsaids (nonsteroidal anti-inflammatory drugs such as aspirin, indomethacin, and ibuprofen)
- Sulfites (commonly used as a preservative for processed foods and beverages)
- Some cooking odors

The severity of an asthma attack can escalate rapidly, so it is important to treat these asthma symptoms immediately once you recognize them.

Without immediate treatment, such as your asthma inhaler or bronchodilator, your breathing will become more labored. If you use a peak flow meter at this time, the reading will probably be greater than 50 percent.

As your lungs continue to tighten, you will be unable to use the peak flow meter at all. Gradually, your lungs will tighten so there is not enough air movement to produce wheezing. This is sometimes called the "silent chest," and it is an ominous sign. You need to be transported to a hospital immediately. Unfortunately, some people interpret the disappearance of wheezing as a sign of improvement and fail to get prompt emergency care.

If you do not receive adequate asthma treatment, you will eventually be unable to speak and will develop a bluish coloring around your lips. This color change, known as cyanosis, means you have less and less oxygen in your blood. Without aggressive treatment for this asthma emergency, you will lose consciousness and eventually die.

If you are experiencing an asthma attack, follow the ***"Red Zone"*** or emergency instructions in your Asthma Action Plan immediately. These symptoms occur in life-threatening asthma attacks. You may need medical attention right away.

Chapter 7: How Widespread Is Asthma In Children?

Asthma is the leading cause of chronic illness in children. It affects as many as 10%-12% of children in the United States alone and, for unknown reasons, is steadily increasing. In the United Kingdom, the amount of children is about the same. Asthma can begin at any age (even in the very elderly), but most children have their first symptoms by the age of five.

There are many risk factors for developing childhood asthma. These include:

- Nasal allergies (hay fever)
- Eczema (allergic skin rash)
- A family history of asthma
- A family history of allergies
- Frequent respiratory infections
- Low birth weight
- Exposure to tobacco smoke before or after birth
- Black or Puerto-Rican ethnicity

- Being raised in a low-income environment

Chapter 8: Why Are So Many Kids Getting Asthma?

No one really knows the exact reasons why more and more children are developing asthma. Asthma is very common in the UK, affecting 5.2 million people. Many of them are kids. It is about the same amount as the population of Scotland and the United States! Some experts suggest that children spend too much time indoors and are exposed to more and more dust, air pollution, and secondhand smoke. Some suspect that children are not exposed to enough childhood illnesses to direct the attention of their immune system to bacteria and viruses.

Chapter 9: How Is Asthma In Children Diagnosed?

Asthma in children can often be diagnosed based on medical history, symptoms, and physical exam. Keep in mind that oftentimes when you take your infant or older child to the doctor with asthma symptoms, the symptoms may be gone by the time the doctor evaluates the child. That is why parents are key in helping the doctor understand the child's signs and symptoms of asthma.

Medical History & Symptom Description: Your child's doctor will be interested in any history of breathing problems you or your child may have had, as well as a family history of asthma, allergies, a skin condition called eczema, or other lung disease. It is important that you describe your child's symptoms -- cough, wheezing, shortness of breath, chest pain or tightness -- in detail, including when and how often these symptoms have been occurring.

Physical Exam: During the physical examination, the doctor will listen to your child's heart and lungs and look for signs of an allergic nose or eyes.

Tests: Many children will also have a chest X-ray and for those ages 6 and older, a simple lung function test called spirometry. Spirometry measures the amount of air in the lungs and how fast it can be exhaled. The results help the doctor determine how severe the asthma is.

Other tests may also be ordered to help identify particular *"asthma triggers"* for your child's asthma. *These tests may include:*

- Allergy skin testing
- Blood tests (IgE or RAST)
- X-rays

These tests are done to determine if sinus infections or gastroesophageal reflux disease (GERD) are complicating asthma. New tests are constantly being developed for asthma. It would be a good idea to ask your doctor about the most current tests and medicines available on the market for asthma.

Chapter 10: How Is Asthma In Children Treated?

Avoiding triggers, using medications, and keeping an eye on daily asthma symptoms are the ways to control asthma in children of all ages. Children with asthma should always be kept away from all sources of smoke. Proper use of medication is the basis of good asthma control.

Based on your child's history and the severity of asthma, his or her doctor will develop a care plan called a written Asthma Action Plan. This plan describes when and how your child should use asthma medications, what to do when asthma gets worse (falls into the yellow or red zones), and when to seek emergency care for your child. Make sure you understand this plan and ask your child's doctor any questions you may have.

Your child's written Asthma Action Plan is important to successfully controlling his or her asthma. Keep it handy to remind you of your child's daily asthma management plan, as well as to guide you when your child develops asthma symptoms. Also, make sure your child's caregiver and/or schoolteacher has a copy of the Asthma Action Plan, so they will know how to treat the child's symptoms if he or she should have an asthma attack away from home.

Chapter 11: How to Treat & Care for Your Asthma

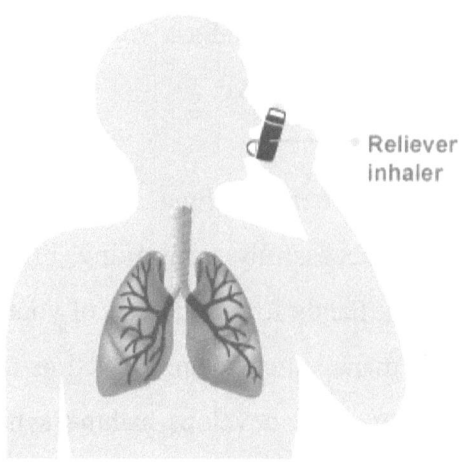

Reliever inhaler

Asthma treatment can vary from anti-inflammatory and bronchodilator asthma inhalers to oral medications to asthma drugs delivered in an asthma nebulizer or breathing machine. Get a better understanding of how asthma medications work so you will know which medications can prevent asthma symptoms. Also, learn about natural asthma remedies and ways to monitor your breathing at home.

Treatment for Asthma

- **Asthma Treatments** - Working together, you and your asthma specialist can design a systematic plan for living with your condition and preventing asthma attacks.

- **Asthma Medications** - Asthma medication can work quickly to stop coughing and wheezing. You will need to work closely with your doctor to determine which asthma medication or combination of asthma medications works best for you. Your age, your symptoms and possible side effects all play a role in choosing the dose and type of asthma medications you need. Asthma medications are different in dose and type for children, women who are pregnant, and for people who have certain health problems or who take certain other medications. Asthma symptoms often change over time. This means you will need to monitor your asthma carefully and work with your doctor to adjust your asthma medications as needed.

- **Asthma Inhalers** - These treatments are used to help asthma. Many people asthma improve with the usage of inhalers. Studies show that the device you use really does not matter, as long as it is used properly. All methods work just as well when the correct technique is used. There are pros and cons to each type of device. Inhalers are smaller and require no power source. Because they deliver the medicine much more quickly than a nebulizer does, some parents may prefer them.

- **Steroids and Other Anti-Inflammatory Drugs** - The most significant treatments for most people with asthma are steroids and other anti-inflammatory drugs. Not only do steroids and anti-inflammatory drugs help people gain better asthma control, but also these asthma drugs help to prevent asthma attacks. Steroids and other anti-inflammatory drugs work by reducing swelling and mucus production in the airways of a person with asthma. As a result, the airways are less sensitive and less likely to react to asthma triggers, allowing people with symptoms of asthma to have better control over their condition.

- **Bronchodilators: Airway Openers** - Almost all people with asthma use one type of medication: a bronchodilator. Short-acting bronchodilators are used only as needed as asthma *"rescue"* medications, while long-acting bronchodilators are used every day to control asthma. Bronchodilators open up the bronchial tubes so that more air can move through. Bronchodilators as well help clear mucus from the lungs. As the airways open, the mucus moves more freely and can be coughed out more easily.

- **Asthma Nebulizer (Breathing Machine)** - Home nebulizer therapy is particularly effective in delivering asthma medications to infants and small children and to anyone who is unable to use asthma inhalers with spacers. The asthma nebulizer changes asthma medication from a liquid to a mist, so that it can be inhaled easily into the lungs. Studies show that the device your child uses really does not matter, as long as it is used properly. All methods work just as well when the correct technique is used.

- **Prednisone and Asthma: Stopping an Asthma Attack** - Sometimes stronger asthma medications are necessary to decrease symptoms. Steroids (anti-inflammatory medications) such as prednisone can be used for asthma as well as other lung diseases. Prednisone and other steroids (inhaled, oral, injection) help calm airway inflammation in asthma. If you have ever had a serious asthma attack; you may have had high doses of steroids in the hospital administered intravenously. When prednisone or other steroid medication is used in a severe asthma attack, no doubt the benefits of steroids prevail over the risks.

Chapter 12: Caring For Asthma Yourself

Some options to help your asthma condition are the following:

- **Asthma Relief** - Treatment of asthma can be broken down into two categories long-term control and quick-relief medications. term control and quick-relief medications.

- Long-term control medications are taken daily to maintain control of persistent asthma. They primarily serve to control airway inflammation.

- The quick-relief medications are taken to achieve prompt reversal of an acute asthma "attack" by relaxing bronchial smooth muscle.

- **Using a Peak Flow Meter** - Have you ever tried a peak flow meter? This asthma test can warn you of an impending asthma attack so you can pretreat before you have serious problems.

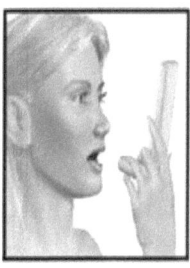  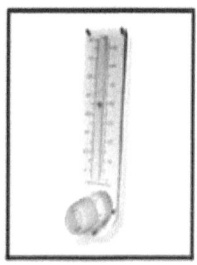

| Take a Deep Breath | Blow out hard and fast | Record the reading on the meter |

- **Developing an Asthma Action Plan** - An asthma action plan is a written plan developed by your doctor or asthma specialist to help you or another family member, including teenagers and children, manage asthma and prevent asthma attacks.

The plan is written to tell you or other family members what to do when there are changes in the severity of asthma symptoms and in peak flow numbers. It may save that person's life!

The following items should be included in your asthma action plan:

- A list of asthma triggers that cause the person with asthma to develop asthma symptoms

- List on how to avoid asthma symptoms

- A list of peak flow meter readings and zones based on the person's personal best reading

- A list of routine asthma symptoms such as coughing, wheezing, tightness in the chest, shortness of breath, and excess mucus production, as well as what you should do if these symptoms occur

- The name and dose of the daily asthma medication that needs to be taken even when you don't have asthma symptoms

- The name and dose of the quick-acting or rescue medication that needs to be taken when you develop asthma symptoms

- The name and dose of the reliever medication that needs to be taken when you are having an asthma attack

- Emergency telephone numbers and locations of emergency care

- Instructions about when to contact the doctor, whom to call if the doctor is unavailable

- **A list of locations** you are able to receive emergency asthma treatment

- **Keeping an Asthma Diary** - Monitoring your asthma is crucial to reduce symptoms. While there is no asthma cure, keeping an asthma diary will help you recognize asthma attacks and prevent them before you become seriously ill.

- **Managing Your Child's Asthma at School** - If your child has been diagnosed with childhood asthma, you know the difficulty of managing your child's asthma symptoms at school. Many children with asthma have symptoms at school. That is why it is important to get the school involved in managing your child's asthma, so a responsible adult at the school knows when and how to administer asthma inhalers or other asthma treatment. Even if your child has mild asthma, working with the classroom teacher and other school officials is vital for managing your child's asthma and treating mild symptoms early on before they worsen. With the prevalence of asthma increasing rapidly among children in the U.S. and United Kingdom, most schools have many students with asthma. Many classroom teachers -- and certainly the school nurses -- are very familiar with helping children with asthma. Still, it is important to take steps to ensure that your child gets adequate attention at school and that all the key people are familiar with what is needed in managing your child's asthma and preventing asthma symptoms at school.

- **Controlling Asthma Triggers** – Triggers are things that make you wheeze or cough. If you or someone you know has allergic symptoms or asthma, you are sensitive to "triggers," including particles carried in the air. These "triggers" can set off a reaction in your lungs and other parts of your body. Triggers are located indoors or outdoors.

They can be simple things like:

- Cold air
- Tobacco smoke
- Wood smoke
- Perfume
- Paint

- Hair spray
- Any strong odors or perfumes

Allergens (particles that cause allergies) such as dust mites, pollen, molds, pollution, and animal dander (which are tiny scales or particles that fall off hair, feathers or skin) from any pets.

Common cold, influenza, and other respiratory illnesses can provoke and cause for asthma to act up.

You may be able to add more triggers to this list. Other things may also trigger your asthma or allergies. It is important to learn which triggers are problems for you. Ask your doctor to help.

Your asthma specialist he/she might suggest:

- Keeping an asthma diary
- Skin testing to test for allergies
- A special diet to look for food allergies

Cold air

Finding triggers is not always easy. If you do know your triggers, cutting down exposure to them may help avoid asthma and allergy attacks. If you do not know your triggers, try to limit your exposure to one suspected trigger at a time. Watch to see if you get better. This may show you if the trigger was a problem for you.

Air Filters and How They Can Help Asthma - When we think of air pollution, we usually associate it with outdoor air. However, with the growing epidemic of asthma in the United States in the last 20 years, especially among infants and children who spend most of their time inside, much attention has been given to indoor air. In fact, in 1990 the United States Environmental Protection Agency (EPA) ranked indoor air pollution as "a high priority public health risk. *The EPA recommends three strategies for reducing indoor air pollution:*

- Controlling sources of pollution,
- Ventilating adequately

- Cleaning indoor air

Holistic Asthma Remedies (The Natural Way) - Looking for some natural asthma remedies? Learn more about herbs, natural dietary supplements, acupuncture, chiropractic, biofeedback, and homeopathy, and how these alternative treatments may alleviate symptoms of asthma.

Curing Asthma Naturally - Looking for a natural asthma cure? Do not try an over-the-counter remedy without reading this information first. This ebook contains some holistic suggestions. You may want to check if it is OK with your doctor first. Everyone reacts differently.

Section Two - Learning How Healthy Living Can Help Your Asthma

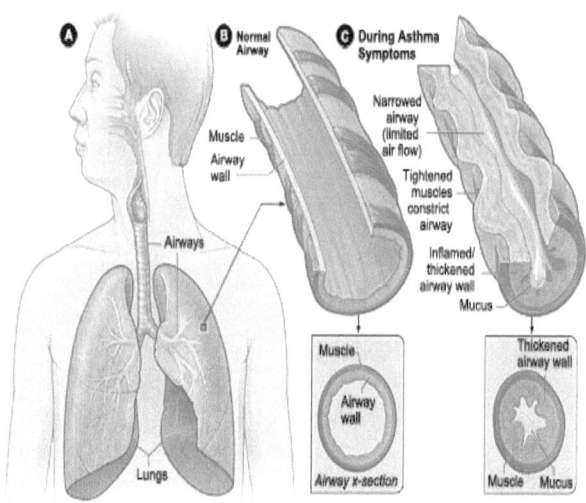

The best way to take care of your asthma is to care for it one day at a time. There is no reason to overwhelm yourself. It is important to breathe well, stay active, and keep asthma symptoms at low. Discover the what diets best for you and try your best to understand your body, so you can control your asthma symptoms, and most important learn how stress and anxiety may trigger an asthma attack.

Chapter 13: Can Your Diet Affect Your Asthma Condition?

There is no special asthma diet. We do not know of any foods that reduce the airway inflammation of asthma. Beverages that contain caffeine provide a slight amount of bronchodilation for an hour or two, but taking a rescue inhaler is much more effective for the temporary relief of asthma symptoms.

However, a good diet is an important part of your overall asthma treatment plan. Just like regular exercise, a healthy diet is good for everyone. That goes for people with asthma, too. Obesity is associated with more severe asthma.

What's more, many doctors suspect that the specific foods you eat might have a direct impact on your asthma. Further research needs to be done before we understand the exact connection between asthma and diet.

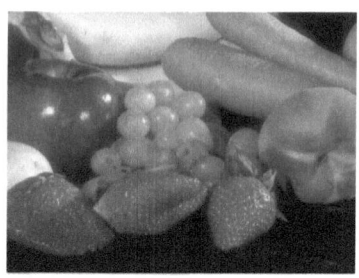

Asthma and Nutrition

The incidence of asthma has risen in the United States during the past three decades, and many researchers believe that our changing diets have something to do with it. As Americans eat fewer and fewer fruits and vegetables and more processed foods, could it be that we are bumping up our risk of developing asthma? Several research studies have suggested this, and others are ongoing, but the connection between diet and asthma remains inconclusive.

There is evidence that people who eat diets higher in vitamins C and E, beta-carotene, flavonoids, magnesium, selenium, and omega-3 fatty acids have lower rates of asthma. Many of these substances are antioxidants, which protect cells from damage.

One recent study of asthma and diet showed that teens with poor nutrition were more likely to have asthma symptoms. Those who didn't get enough fruits and foods with vitamins C and E, and omega-3 fatty acids were the most likely to have poor lung function. A 2007 study showed that children who grew up eating a Mediterranean diet -- high in nuts and fruits like grapes, apples, and tomatoes -- were less likely to have asthma-like symptoms.

However, it is not at all clear that deficiencies of these nutrients actually caused the asthma. Studies that have used specific vitamins and minerals to treat asthma have been unsuccessful. Why? Some researchers think that it might be the interaction of different vitamins, minerals, and other antioxidants that naturally occur in foods that have the health benefits. Therefore, it is unlikely that taking vitamins, minerals, or other food supplements will improve your asthma control and prevent symptoms of asthma.

Regardless of the specific link between asthma and diet, we do know that good nutrition is important for anyone, and especially people with chronic diseases. If you are not getting the right nutrients, your body may be more susceptible to illness and have a harder time fighting the respiratory viruses that often trigger an asthma attack or severe asthma emergency.

What Should I Eat to Help My Asthma?

Given the murky evidence for a link between asthma and nutrition, *here are some foods to put on your plate that are good for asthma:*

- **Eat plenty of fruits and vegetables.** We still do not know which fruits and vegetables might have an effect on asthma, so the best advice is to increase your intake of a wide variety of them. Frankly, we should all eat more of them anyway.

- **Eat foods with omega-3 fatty acids.** Omega-3 fatty acids -- found in fish like salmon, tuna, and sardines and some plant sources, like flaxseed -- are believed to have a number of health benefits. Although the evidence that they help with asthma is not clear, it is still a good idea to include them in your diet.

- Avoid trans fats and omega-6 fatty acids. There is some evidence that eating omega-6 fats and trans fats, found in some margarines and processed foods, may worsen asthma. Since they raise your risk of cardiovascular disease anyway, this is just another reason to reduce the amount you eat.

Will this dietary advice help your asthma symptoms? Medical doctors are uncertain. We cannot be sure. But while we may not know what specific vitamins, minerals, and antioxidants might help with asthma, if any, there's no downside to eating more fruits, vegetables, and (in moderation) healthy fats.

What Foods Can Affect My Asthma Symptoms?

- **Diets High in Calories:** If you eat more calories than you do burn, you will gain weight. That is bad not only for your general health, but for your asthma specifically. People who are obese are more likely to have more severe symptoms, take more medication, and miss more work than people who maintain a normal weight.

- **Food Allergies:** Many people have food intolerances (such as lactose intolerance), but these are not true allergies and rarely worsen asthma. Only about 2% of adults with asthma have true food allergies to milk, eggs, shellfish, peanuts, or other foods. When exposed to even small amounts of the foods to which they have become allergic, these patients can have life-threatening anaphylactic attacks, including bronchospasm, which requires immediate asthma medication. There is no evidence that elimination of all dairy products from the diet improves asthma control, even in a minority of patients. That is simply "an old wives tale" and can lead to osteoporosis, especially in patients who must regularly take corticosteroids to control their severe asthma.

- **Preservative Sensitivities:** Sulfites, which are used to keep food fresh and stop the growth of mold, can trigger temporary asthma symptoms in a few people with asthma. Sulfites can give off sulfur dioxide that can irritate the lungs. Sulfites are no longer added to fresh fruits and vegetables in the U.S. Nevertheless, they are still used in many processed foods, and may be in condiments, dried fruits, canned vegetables, wine, and other foods.

- **Gastroesophageal Reflux Disorder (GERD):** Up to 70% of all people with asthma also have GERD (reflux of stomach acid), which can make asthma more difficult to control. Sometimes GERD does not cause typical heartburn symptoms. If you have GERD, you may need to take medicine. Weight loss is often all that is necessary to eliminate GERD. You should also try eating smaller meals and cutting down on alcohol, caffeine, and any foods that you notice trigger GERD symptoms. Avoid eating just before bedtime.

Before you make any big changes to your eating habits, it is always a good idea to talk to your doctor or asthma specialist first. Depending on your asthma diagnoses and considering your general health and the severity of your asthma symptoms -- your doctor might have specific advice for how to improve your diet.

Chapter 14: The Holistic Approach to Treating an Asthma Condition

With asthma, airways of the lungs, called bronchial tubes, are inflamed. The surrounding muscles constrict and mucus is produced, which both cause airways to narrow. Untreated asthma flare-ups can lead to hospitalization and can even be fatal. Asthma is not a condition that should be self-treated. A doctor's supervision is required.

A Russian-born researcher Konstantin Pavlovich Buteyko developed the Buteyko (pronounced bew-

TAY-ko) Breathing Technique. It consists of shallow-breathing exercises designed to help people with asthma breathe easier.

The Buteyko Breathing Technique is based on the premise that raising blood levels of carbon dioxide through shallow breathing can treat asthma. Carbon dioxide is believed to dilate the smooth muscles of the airways. A study involving 60 people with asthma compared the effects of the Buteyko Breathing Technique, a device that mimics pranayama (a yoga breathing technique), and a placebo. Researchers found people using the Buteyko Breathing Technique had a reduction in asthma symptoms. Symptoms did not change in the pranayama and the placebo groups.

The use of inhalers was reduced in the Buteyko group by two puffs a day at six months, but there was no change in the other two groups.

There have been several other promising clinical trials evaluating this technique.

Omega Fatty Acids

The primary inflammation-causing fat in our diets is called arachidonic acid. A German study examined data from 524 children and found that asthma was more prevalent in children with high levels of arachidonic acid. Arachidonic acid is found in food, particularly egg yolks, shellfish, and meat. Eating less of these foods may decrease inflammation and asthma symptoms.

Arachidonic acid can also be produced in our bodies. Another strategy to reduce levels of arachidonic acid is to increase intake of beneficial fats such as EPA (eicosapentanoic acid) from fish oil, and GLA (gamma-linolenic acid) from borage or evening primrose oil.

Fruits and Vegetables

A study examining food diaries of 68,535 women found that women who had a greater intake of tomatoes, carrots, and leafy vegetables had a lower prevalence of asthma.

High consumption of apples may protect against asthma.

Daily intake of fruits and vegetables in childhood decreased the risk of asthma. A University of Cambridge study found that asthma symptoms in adults are associated with a low dietary intake of fruit, vitamin C, and manganese.

Butterbur

Butterbur is a perennial shrub that grows in Europe, Asia and North America. The active constituents are petasin and isopetasin, which are believed to reduce smooth muscle spasm and have an anti-inflammatory effect.

Researchers at the University of Dundee, Scotland, evaluated the effects of the herb butterbur in people with allergic asthma who were also using inhalers. They found that butterbur added to the anti-inflammatory effect of the inhalers.

Another study examined the use of butterbur root extract in 80 people with asthma for four months.

The number, duration, and severity of asthma attacks decreased and symptoms improved after using butterbur. More than 40 percent of people using asthma medication at the start of the study reduced their intake of medication by the end of the study.

The butterbur plant contains pyrrolizidine alkaloids, which can cause liver damage. Only extracts in which the pyrrolizidine alkaloids have been removed should be used.

Bromelain

Bromelain is an extract from pineapples that is believed to be a natural anti-inflammatory. Researchers at the University of Connecticut found that bromelain reduced airway inflammation in animals with allergic airway disease. Side effects may include allergic reactions in some people.

Boswellia

The herb boswellia, known in Indian Ayurvedic medicine as Salai guggal, has been found to inhibit the formation of compounds called leukotrienes.

Leukotrienes released in the lungs cause narrowing of airways. A double blind, placebo-controlled study of forty patients, 40 people with asthma were treated with a boswellia extract three times a day for six weeks. At the end of this time, 70 percent of people had improved.

Numerous studies have found that obesity is a risk factor for asthma. Biofeedback. Biofeedback has long been recommended as a natural therapy for asthma.

Chapter 15: Asthma & Exercise

One of the goals of asthma treatment is to help you maintain a normal and healthy lifestyle, which includes exercise and other physical activities. Following your asthma action plan by taking medications as prescribed by your doctor, avoiding triggers and monitoring your symptoms and lung function will help you achieve this goal.

If asthma symptoms prevent you from participating fully in activities, talk to your asthma doctor. A small change in your asthma action plan may be all that is needed to provide asthma relief during exercise or activity.

What Types of Exercise Are Best for People With Asthma?

Activities that involve short, intermittent periods of exertion such as volleyball, gymnastics, baseball and wrestling are generally well tolerated by people with symptoms of asthma.

Activities that involve long periods of exertion are:

- Soccer
- Distance running
- Basketball
- Field hockey

These sports may be less well tolerated. Also less well tolerated are cold weather sports such as ice hockey, cross-country skiing, and ice-skating. However, many people with asthma are able to fully participate in these activities.

Swimming, which is a strong endurance sport, is generally well tolerated by many people with asthma because it is usually performed in a warm, moist air environment. It is also an excellent activity for maintaining physical fitness.

Other beneficial activities for people with asthma include both outdoor and indoor biking, aerobics, walking, or running on a treadmill.

How Much Exercise Should I Get?
Generally, exercise should be performed 4-5 times per week for at least 30 minutes. Talk to your doctor to find out how much exercise you should get

What Should I Do to Control My Asthma When I Exercise?

- Always use your pre-exercise asthma inhalers (inhaled bronchodilators) before beginning exercise

- Perform warm-up exercises and maintain an appropriate cool-down period after exercise

- If the weather is cold, exercise indoors or wear a mask or scarf over your nose and mouth

- If you have allergic asthma, avoid exercising outdoors. Especially, when pollen counts are high and when there is high air pollution

- Restrict exercise when you have a viral infection, like a cold

- Exercise at a level that is appropriate for you

Maintaining an active lifestyle is important for both physical and mental health. Remember: asthma is not a reason to avoid exercise. With proper asthma diagnosis and the most effective asthma treatment, you should be able to enjoy the benefits of an exercise program without experiencing asthma symptoms.

What Do I Do If I Have an Asthma Attack While Exercising?

If you begin to experience asthma symptoms during exercise, stop and repeat your pre-exercise inhaled medication (quick relief medication like albuterol). If your symptoms completely go away, you may restart the exercise. If your symptoms return, stop the activity, repeat your quick relief medication and call your health care provider for further advice.

Section Three – How Your Home Affects Your
Asthma

Chapter 16: Why Are So Many People Being Diagnosed With Asthma?

Are one of the reasons so many people are developing asthma is because of the cleaning products we use?

In the last ten years, many of my friends and their family have been diagnosed with asthma. I ask you why is asthma on a rise? Why are many people developing asthma? There are at least three or four kids every year in my children's elementary classes who suffer from asthma.

Why am I telling you this because I want you to understand that it does not matter what age; you can develop asthma at any age.

Now, the big question, "Why?" I will tell you.
The air inside your home is an extension of your lungs. You eat approximately two to three pounds of food per day, drink about three pounds of liquid, and breathe 15 pounds of air. You can live 40 days

without food, three days without water, but only three minutes without air.

Whatever you eat passes through your digestive system enabling your body to separate nutrients and waste material. Your lungs are not capable of doing this type of function. They have no such defensive system.

What you breathe goes directly into the bloodstream. It is carried by the use of the blood to every cell in the body.

Research from organizations have stated that the poor air quality in our homes is so serious, that this maybe one of the reasons for the rise of asthma in our society. Our indoor air quality is the most significant environmental issue, which we have to face now and into the next decade.

Research has stated that the worst air pollution is inside our homes. A five-year study, done by the EPA, showed that many homes had chemical levels that were 70 times higher than that of the air outside. In fact, they reported that cleaning and personal care products, commonly found in every home are three times more likely to cause asthma than air-born pollutants. It is estimated that 1,500 hazardous substances find their way into a typical home.

Top environmental organizations have stated that their studies concluded thousands of people get asthma each year and it is caused by indoor air pollutants. A scientific study was also done that revealed that because of homemakers has a 55% higher risk of asthma than men and women working outside the home.

The tragedy is that children are the most susceptible to this toxic environment. Children's respiratory rates are three times higher than adults are; therefore, they inhale and absorb three times the amount of toxic cleaning supplies. Maybe this is why asthma in children is on a rise also in many countries such as the United Kingdom and the United States.

Our home, which ought to be a haven of safety for our children, has become a dangerous environment. Since 1960, there has been an 80% increase in respiratory problems amongst children. For many of these children, inhalers have become an enduring part of their lives. A temporary protection against foreseeable tragedy!

We go to the store to purchase cleaning supplies with the notion that their safe and they're going to kill the germs in our home and make our home sparkle.   When we go to the store to get our cleaning supplies we buy our laundry detergent, shampoos, and dish soap, believing them to be safe; not realizing that we are bring toxic poisons into our homes that could be deadly in the long-run.

Formaldehyde is in almost everything you use in your home, from toothpaste to laundry soap. It is used as a preservative. Nine billion pounds are produced every year. Medical organizations recently told the pubic media that there is such a high degree of formaldehyde in our bodies that, when we die, we no longer decay.

The National Institute of Occupational Safety and Health warned the public that formaldehyde should be handled with caution. It is suspected of causing many different medical issues **such as:**

- Birth defects
- Genetic damage
- Headaches
- Joint pain
- Chest pains
- Depression
- Ear infections
- Chronic fatigue
- Dizziness
- Loss of sleep

Here are some examples of cleaners commonly found in homes.

Dishwashing Detergent

Dishwashing detergents have caused more household poisonings than any other cleaning product in the home. Nearly all dishwashing detergents contain naphtha, a fuel used in camping stoves. Naphtha is a central nervous system depressant. Other high-tech cleaning agents included are diethanolsamine, which is a liver poison, and chlorophenylphenol, which is a toxic metabolic stimulant. Chlorine is a poison present in nearly all dishwashing detergents. When washing your dishes, these chemicals are being released into your breathing space.

Air Fresheners

Were you aware that air fresheners can damage nerves or they can coat your mucus membranes with an oily film, hindering your natural ability to smell? The fresh air in a can may be able to fool humans but you will notice your pets wisely running for cover.

Tub and Tile Cleaners

On the back of Tile Power, there is a warning, Use in well-ventilated areas. When have you been in a well-ventilated bathroom? This product is not recommended for people with heart conditions or chronic respiratory problems. They go on to warn, avoid prolonged inhalation of fumes. Yet, on the front of the container, in bold attractive lettering, it proclaims, NEW FRESH SCENT. While enjoying the new fresh scent, you are being killed by the sodium hypochlorite and sodium hydroxide. Tile cleaners commonly contain ammonia and ethanol. Many people have complained of dizziness and nausea in using these powerful chemically spiked cleaners.

Laundry Soap

What is on your clothes goes into your skin. It has a known fact. Skin is a semipermeable membrane and is the largest organ of the body. Nicotine patches reveal how absorbent our skin is. If it is on your skin, it is in your bloodstream. Phosphorus, ammonia, napthalcaene and phenol, which are deadly poisons, are found in most laundry soaps. Some companies have proclaimed their products environmentally safe because they use enzymes. Enzymes have been considered a miracle-cleaning agent. They direct the natural, cleaning ability of water to protein. This enables water to suspend dirt with great effectiveness. Yet enzymes are unable to detect a dirt protein from the protein in your skin and mucus membranes. Serious inflammation can result from breathing enzyme particles into the nasal passages and lungs. Clothes washed in enzymes can cause inflamed spots or rashes on the skin.

Fabric Softeners

Fabric softeners use an oily residue to cut down on static cling. Because it is not washed or rinsed from the clothes, a high concentration of toxins remains next to your skin and is being continuously absorbed into the bloodstream. Ammonia propellants and powerfully, strong synthetic fragrances can cause irritation, stuffy noses and watery eyes.

Oven Cleaners

I cannot tell you how many times I had to leave the house after using an oven cleaner because I started choking and coughing like crazy. Head in oven, holding our breath, squinting because of the fumes, we spray a toxic film where we cook supper. Holding our breath is no solution since the toxic vapors will linger in the house for over three months. Bet you did not know they stood in your homes for that long. Imagine what is going into our food when we cook. Until recently, there has been little choice in improving the air quality of the home. There are now natural cleaning products that produce safe, environmentally friendly cleaning products. This is why global warming has become such a major issue. We are destroying the planet and we are killing ourselves. Change is good. It is time to change our cleaning products. It is time to use natural cleaning products.

Chapter 17: Why People Need To Clean With Natural Cleaning Supplies

Nowadays everyone is suffering from allergies. I know at least 10 people of the top of my head whose kids suffer from asthma. Did you ever ask yourself why? Why do so many people suffer allergies, medical conditions, and various illnesses? Could it stem from the cleaning products and the chemicals we use inside and outside our home?

How healthy is your home? Your home may sparkle because it is clean, but in the end, those chemicals might be damaging you and your family's bodies more than you are aware.

The problem is in today's world everyone is always on the run. Your either working, running to school events, running to after school activities, chauffeuring the kids to wherever, and tending to daily responsibilities. All these responsibilities fill up your daily schedules. With all this going on in your busy schedule honestly who has time to research how hazardous each cleaning product might be. Who has time to examine each chemical in every product and the long-term effects it might have on people and animals. How many people have the time to seriously think about what cleaning products are safest to use in the home? Who has the time to research this topic come on now?

But if you really want to protect the one's you love than the health of your home is crucially important and this is why you need to know what products you are using to clean them. Even though you're busy, make time, your family health comes first. Nothing matters more than a healthy home with happy people living in that home. Below are 3 reasons why you should be cleaning your home with natural cleaning products.

Do it for yourself, your family and the environment:

1. **Safer & healthier:** Do you know what is in the household cleaning products you are using? The fact is, most household cleaning products on market shelves today contain many harmful and hazardous toxins and chemicals. These hazardous toxic substances can enter the body through ingestion, inhalation or absorption through the skin, which can affect the heart, lungs, kidneys, liver and the brain and have the potential to cause many serious health problems. Natural cleaning products on the other hand are non-toxic and much safer to use because they are made from natural, wholesome, ingredients. In addition, when it comes to cleaning and disinfecting your home, they work just as effectively if not better than the chemical products that sell in thousands of stores.

2. **Better for the environment:** Most retail brand household cleaning products are made from harmful chemicals and these

substances later become pollutants for our air and our landfills. Natural cleaning products on the other hand are made from natural ingredients, which are non-toxic as well as biodegradable and thus have a reduced impact on our environment. In addition, most natural cleaning products are frequently made from recycled packaging and materials.

3. **Save you money:** It should come as no surprise to you that the average costs for keeping up a family and a household are continuing to rise. However, you do not have to endanger the health of your families or your homes by using expensive, toxic chemical cleaning products. Natural cleaning products are not only safer to use, they also save you money because they last longer and cost less. In addition, to have a pretty scent in your home add 100% natural essential oils to your home made recipes such as tea tree oil, lavender or lemon oil which not only provide a fresh aroma and

scent, but also have natural antiseptic, antifungal and disinfectant properties contained within them.

Chapter 18: Asthma Proofing Your Home

If your child has asthma, you want to create the best home environment possible. To do that, you need to know which things are triggers for your child and then take steps to eliminate them - as best you can - from your home.

During the winter, everyone usually remains inside this time of year. Indoor air quality becomes poor, while colds and flu's affect you families, the people in your children's schools and communities. As homes are tightly sealed with closed windows, air is trapped in the home with less ventilation causing irritants that cause asthma to be more concentrated.

Tobacco smoke, molds, smoke from festive holiday fireplaces and wood stoves used for heat are part of the cause to poor air quality in your home. When these things occur each winter it negatively affects asthma sufferers making their conditions worsen. Asthma is on a rise. Children are increasingly diagnosed with asthma in the United Kingdom and the United States each year. Asthma sufferers can be especially affected by air quality in their environment.

Asthma sufferers often have spikes in asthma symptoms following respiratory symptoms associated with colds and flu. This is the primary reason for increased asthma attacks in the winter. The higher incidence of respiratory illness during this season means this time of year can be twice as hard for those with asthma because colds, flu and other infections like pneumonia can trigger symptoms. Cold or dry winter air can also be a factor. An allergy to dust, pets, animals, pollens or molds also leads to asthma in some individuals.

The number one reason for hospital asthma admissions is colds, even among those with allergy-mediated asthma. However, there are things families can do to help children with asthma get through this time of year.

Diligent hand washing to help prevent colds and keeping air in the home as clean as possible goes a long way. It is also important to lessen the particular allergens that are triggers for each child.

A research study was conducted in 2007. The study's results showed that changes in the home environment could produce reduction in symptoms comparable to that achieved with asthma medications.

New asthma guidelines released in August of 2007 suggest that reducing dust mites has the strongest impact of all environmental strategies for reducing asthma triggers in homes.

One in every eight children worldwide has asthma. Children-of-color, those who live in the inner city, and those from lower-income families in both urban and rural areas are at greater risk for developing asthma. Increased asthma incidence in these populations may stem in part from higher exposure to multiple indoor allergens and tobacco smoke.

Asthma is a chronic condition of the passageways that carry air to the lungs. These airways become narrow and their linings become swollen, irritated and inflamed. Bronchial constriction, or tightening of the smooth muscles that surround these airways, is the feature of asthma that typically leads to "asthma attacks," which can lead to hospitalization in some children. Below are tips created by asthma specialists, these tips were created to teach you how to asthma proof your home.

Tips to help "asthma-proof" your home:

1. Keep children away from second-hand tobacco smoke. If anyone in your household smokes, make sure, they smoke outside or far away from children. Do not allow

anyone to smoke in the car when children are present.

2. Wood stoves and fireplaces have to be vented well outside, with good draft and well-functioning chimneys. Only light fires when local air quality is good, and only burn wood with enough paper and kindling to start the fire. Never burn other items. Keep as much smoke out of the room as possible by closing fireplace doors.

3. Use only clean-burning candles. Plain, unscented natural beeswax candles have a gentle, sweet scent without added perfumes and dyes.

4. Avoid bedding or pillows stuffed with down, feathers or foam rubber. Cover pillows and mattress with dust-proof, "hypoallergenic" allergen-impermeable fabric cases. Avoid lots of stuffed animals and plush toys.

5. Wash bedding once a week in hot water, and dry in a hot dryer for at least 30 minutes to eliminate dust mites.

6. Dust and vacuum when your child is out of the room. Use a damp cloth for dusting, and a vacuum that collects and traps dust mites such as one with a double bag or HEPA (high-energy efficiency particulate air) filter.

7. Keep pets off furniture and out of your child's room.

8. Use HEPA filters in key locations in your home, especially in your child's bedroom.

9. Rid your home of cockroaches and other pests that may carry allergens. **To avoid cockroaches or any other types of bugs:** Have your home professionally exterminated every few months. In between professional treatments, use bait traps to catch roaches (avoid aerosol sprays, which can aggravate asthma).

o Avoid saving boxes, paper bags, or newspapers in piles around your home.

o Do not leave open food or dirty dishes lying around your kitchen.

o Keep counters free of crumbs or spills.

o Keep garbage containers closed.

o Wash recyclables before putting them in the bin.

10. Keep your child's medicines including asthma inhalers on hand and up-to-date. Ask your doctor to check your child's lung function and sensitivities to airborne allergens. Also, ask them to help you make a written asthma management plan for your child.

To help the asthma sufferer in your home you will need to get rid of any mold or mildew in your home. Mold is a big factor for people with asthma. By reducing or eliminating the mold or mildew in your home, you can help your asthma tremendously. The steps below will help you rid mold or mildew from your home.

Molds are microscopic plant-like organisms. They can grow on many surfaces and flourish in damp places like bathrooms and basements. Molds reproduce by sending spores into the air; inhaled mold spores are a common asthma trigger.

To reduce moisture and mold:

- Fix leaky pipes, faucets, or roofs. Clean and repair roof gutters regularly.

- Make sure your bathrooms and basement are ventilated well. Install and use exhaust fans to help lower moisture in these areas.

- If you have any damp closets, clean them thoroughly and leave a 100-watt bulb on all the time to increase the temperature and dry out the air.

- Run a dehumidifier in the basement or other damp areas. Again, it is important to empty and clean the water pan often.

- Remove wallpaper and wall-to-wall carpeting from bathrooms and basement rooms.

- Run the air conditioning (this is especially helpful if you have central air), making sure to change the filter monthly.

- Avoid houseplants, which may harbor mold in their soil.

- Clean any visible mold or mildew with a solution that is one part chlorine bleach to 10 parts water. Do not paint or caulk over moldy surfaces without cleaning them first.

- When painting bathrooms or other damp areas of your house, use anti-mildew paint.

- If there is visible mold on ceiling tiles, remove and replace them. Also, check to see if there is a leaky pipe that may be causing the problem.

- Replace or wash moldy shower curtains.

Chapter 19: Asthma & Dust – The Relation Between The Two

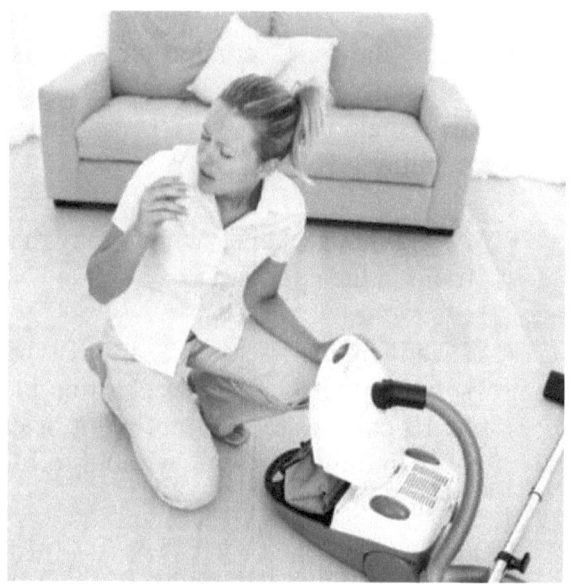

Asthma is so common in our society, that we accept it as 'one of those things', particularly when the doctor tells us just to keep taking the medicine and the nebulizer.

Could dust be the main cause of your asthma or your child's asthma? Nevertheless, what if you could find out the cause that's triggers your the asthma condition? Could you do something about it and reduce or end your dependency on the asthma medicines you have to take?

Many doctors and asthma specialists believe that dust is one of the main causes of asthma. They say the dust mites in your home trigger the asthma attacks. You are probably wondering what are dust mites? They sound disgusting. Well they are disgusting.

What do I do first?

1. Start in the bedroom. Most of the dust mites in your house live in your mattress.

2. Put a tightly woven, dust-proof cover over your mattress.

3. Wash your sheets and blankets in very hot water every week.

4. Wash your pillow every week or put a dust-proof cover on it. The pillowcase goes over the cover.

5. The water used to wash your sheets and blankets should be 130°F to 140°F. This temperature is higher than you may want for your water heater, because water over 120°F can burn children if they turn on the hot water by themselves. If you do not want to set your water heater at this temperature, you can wash your sheets and blankets at commercial laundries.

6. Your bedroom should have a hardwood, tile or linoleum floor instead of carpet. Dust mites can grow rapidly in carpet. If you must use carpet, try not to place it on concrete because the warm space between a rug and concrete is a good place for mites to live.

7. Vacuum your carpets and upholstery every week. Vacuums with high-efficiency filters pick up more dust mites, but even standard vacuums will help.

8. Plastic or wood furniture that does not have much padding can also help keep down the number of dust mites in your home. Because dust mites love warm, humid places, keep the humidity in your home low by using a dehumidifier and running your air conditioner. Special air filters can also help reduce dust mites in the air.

9. Use a damp cloth or rag weekly to wipe surfaces where dust can collect. This includes countertops, shelves and windowsills. If your children have allergies, make sure to buy them stuffed animals that you can put in the washing machine.

Section Four - Coping With Asthma Emotionally

Chapter 20: Asthma & Stress

Although stress does not cause asthma, stress and asthma are definitely linked. Asthma causes stress, and stress makes it more difficult to control asthma. Even daily stress can make your asthma symptoms worsen.

Learning to change your stress response to decrease your asthma symptoms is important. Equally important is prioritizing your daily schedule so you allow enough time to accomplish what you need to do without feeling pressured or overwhelmed.

The longer breathing problems go uncontrolled, the more likely you will notice the signs caused by stress. This can make it more difficult to breathe and create even further problems, *including:*

- Difficulty sleeping and Nocturnal asthma causing constant fatigue

- Inability to exercise or exercise-induced asthma, leading to poor aerobic and physical fitness

- Difficulty concentrating, leading to poor performance

- Increased irritability from lack of sleep or asthma medication side effects

- Withdrawal from favorite activities because of lack of energy

- Changes in appetite because of medications

- Feelings of depression

There is a better way to live with asthma and prevent asthma symptoms. Learn all about stress and your stress response. Set goals to manage your stress in a way that is healthy and not detrimental to your breathing.

Stress is a common asthma trigger. An asthma trigger is anything that brings on asthma symptoms. When you have stress and asthma, you might feel short of breath, anxious, and even panicked. Stress may cause your asthma symptoms to worsen and cause you to feel frightened.

When stress levels start to creep upward -- whether it is over bills, work, or your kids' jam-packed calendar -- asthma symptoms can kick into overdrive. As the wheezing and coughing gets worse, your health becomes one more reason to worry. Asthma, stress, and anxiety make for a vicious circle, and one that can spiral downward quickly.

When Asthma Treatment Triggers More Anxiety

With persistent asthma, you have symptoms more than once a week, but not constantly. Treating persistent asthma requires long-term maintenance therapy, such as an inhaled steroid, plus rescue therapy when something triggers symptoms. In addition, when your symptoms are out of control (in the red zone, a severe asthma attack), prednisone for asthma might be necessary for a few days. The problem is that prednisone often causes mood swings as a side effect, adding fuel to the anxiety fire.

Remember, prednisone is a short-term treatment for most people with asthma. After you finish taking the "burst" of oral steroids, your mood will return to normal. Inhaled steroids do not cause permanent mood changes.

If your long-term asthma medication does not work well, and wheezing and chest tightness occur too often, a vicious circle can begin where anxiety worsens asthma, and asthma worsens anxiety. That is when you need to talk to your doctor about your symptoms, triggers, and stress. Also, discuss other asthma treatment options that can get your asthma under control again, so you can prevent symptoms of asthma.

How to Manage Stress with Asthma

- **Stress is part of daily life:** With or without asthma. That is why it is important to find effective ways to manage stress with asthma. Learning to relax before you feel stressed can help you prevent shortness of breath and avoid an asthma attack.

- **Change Your Thoughts:** Learn to change thought patterns that produce stress. What you think, how you think, what you expect, and what you tell yourself often determine how you feel and how well you manage rising stress levels.

- **Reduce Your Stressors:** To reduce stressors, you must identify the major stressors in your life such as money problems, relationship problems, grief, too many deadlines, and lack of support. If you cannot resolve these stressors alone, get professional help for problems that are too difficult to deal with by yourself.

- **Avoid Stressful Situations**: Try to avoid situations that trigger stress for you. Practice effective time-management skills, such as delegating when appropriate, setting priorities, pacing yourself, and taking time out for you.

- **Exercise Daily:** Get some exercise. Exercising with asthma is an excellent way to burn off the accumulated effects of stress and keep your body healthy.

- **Get Plenty of Sleep:** With asthma or any chronic illness, you need plenty of sleep. If you are not sleeping well or suffer with nighttime asthma, you will have less energy and fewer resources for coping with stress.

Developing good sleep habits is very important. *Here are seven sleep tips:*

1. Do not go to bed until you are tired.

2. Develop specific bedtime rituals and stick to them.

3. If you have trouble sleeping, do not watch TV, read, or eat in bed.

4. Do not engage in exercise or strenuous activity immediately before bedtime.

5. Avoid caffeine

6. Do not nap

7. Go to bed and get up at the same time every day, including on weekends.

Eat a Healthy Diet. Junk food and refined sugars low in nutritional value and high in calories can leave you feeling out of energy and sluggish. Limiting sugar, caffeine, and alcohol can promote health and reduce stress.

Delegate Responsibility. Stress often results from having too many responsibilities. You can free up time and decrease stress by delegating responsibilities. Take a team approach and involve everyone in sharing the load.

Try using these eight guidelines at home or modifying them to fit your situation at work:

1. Make a list of the types of tasks involved in the job.

2. Take time to train someone to do the job or specific tasks.

3. Assign responsibility to a specific person.

4. Rotate unpleasant duties.

5. Give clear, specific instructions with deadlines.

6. Be appreciative; let people know you are pleased by a job well done.

7. Allow others to do a job their own way.

8. Give up being a perfectionist.

Seek Support. Life is tough sometimes and support from friends and family members is important. In fact, social support is the single most important cushion/shield against stress. Here are some tips you can offer to your family or friends when they ask you how they can help.

Family and friends can do the following:

1. Help you remain as active and independent as possible.

2. Provide emotional support.

3. Help with household chores and with grocery shopping and other errands as necessary.

4. Learn what they can about your condition and prescribed treatment by attending doctors' appointments with you.

5. Provide encouragement and help you follow your prescribed asthma treatment plan.

Practice Relaxation Exercises

Relaxation exercises combine deep breathing, releasing of muscle tension, and clearing of negative thoughts. If you practice these exercises regularly, you can use relaxation exercises when needed to lessen the negative effects of stress.

Relaxation exercises include diaphragmatic and pursed lip breathing, imagery, repetitive phrases (repeating a phrase that triggers a physical relaxation, such as "relax and let go"), and progressive muscle relaxation. Many commercial audiotapes, CDs, and books that teach these exercises are available.

Relaxation Exercises to Manage Stress with Asthma

This is a two-minute Relaxation Exercise. Concentrate your thoughts on yourself and your breathing. Take a few deep breaths, exhaling slowly. Mentally scan your body. Notice areas that feel tense or cramped. Quickly loosen up these areas. Let go of as much tension as you can. Rotate your head in a smooth, circular motion once or twice. (Stop any movements that cause pain.) Roll your shoulders forward and backward several times. Let all of your muscles completely relax. Recall a pleasant thought for a few seconds. Take another deep breath and exhale slowly. You should feel more relaxed.

Mind Relaxation Exercises

Close your eyes. Breathe normally through your nose. As you exhale, silently say to yourself the word "one," a short word such as "peaceful," or a short phrase such as "I feel quiet" or "I'm safe." Continue for 10 minutes. If your mind wanders, gently remind yourself to think about your breathing and your chosen word or phrase. Let your breathing become slow and steady.

Deep Breathing Relaxation

Imagine a spot just below your navel. Breathe into that spot and fill your abdomen with air. Let the air fill you from the abdomen up, and then let it out, similar to deflating a balloon. With every long, slow breath out, you should feel more relaxed.

For people who want to help themselves relax, but are unfamiliar with all the different relation techniques I am going to show you techniques that you can practice in public or the privacy of your own home. This exercise will definitely relax you and help you from getting asthma attacks.

First, you need to need to learn how to incorporate daily relaxation steps in to your daily routine that are simple, quick, and easy to do.

When you feel stressed, you are probably thinking to yourself, "I wish I knew some relaxation techniques that I could do to help me relax, and also, relieve all my stress and tensions in my daily life.

Now it's time with the techniques listed below:

First, no matter what your daily responsibilities are everyone needs to take small breaks to give yourself a breather. Small breaks will re-energize you and help you become more focus. You cannot do your best if you are not focused.

Using a few yoga poses, you can increase body awareness and prevent physical problems- such as your asthma condition

Practice these yoga positions at school, work, or at home to help you recover from that terrible illness called "stress."

Tadasana (Mountain Pose)

In many series of Yoga Exercises, Tadasana is a position used at the beginning, in the middle, and in the end, in which you pay attention to your position, your concentration and your breathing. During intensive Yoga sessions, Tadasana makes it easier for you to maintain your meditative focus, as well as to increase and regain it.

1. Stand up straight with both feet at hip-width.

2. Turn your heels a little outward and let your weight rest on your toes.

3. Your arms hang downwards along your body and the palm of your hands point towards your body.

4. Now make the back of your pelvis move away from your lower back. You can do this by drawing in your ribs a little in the direction of your belly.

5. Breathe in and out a few times with full concentration. Through your breathing, place your neck straight over the upper back. It would then feel as if you stretch your body upwards from the neck.

6. The shoulders feel broad and are relaxed.

7. Your breathing is free and relaxed.

8. Look straight ahead of you at a spot within your vision and try to stand motionless with as little effort as possible.

9. Whenever you do this exercise, do it with care and always try to increase your focus and your relaxation.

Garundasana (Eagle Pose)

The word is usually translated into English as "eagle," though according to one dictionary the name literally means "devourer," because Garuda was originally identified with the "all-consuming fire of the sun's rays."

Steps

1. Stand, bend your knees slightly, lift your left foot up and, balancing on your right foot, cross your left thigh over the right. Point your left toes toward the floor, press the foot back, and then hook the top of the foot behind the lower right calf. Balance on the right foot.

2. Stretch your arms straightforward, parallel to the floor, and spread your scapulas wide across the back of your torso. Cross the arms in front of your torso so that the right arm is above the left, and then bend your elbows. Snug the right elbow into the crook of the left, and raise the forearms perpendicular to the floor. The backs of your hands should be facing each other.

3. Press the right hand to the right and the left hand to the left, so that the palms are now facing each other. The thumb of the right hand should pass

in front of the little finger of the left. Now press the palms together (as much as is possible for you), lift your elbows up, and stretch the fingers toward the ceiling.

4. Stay for 15 to 30 seconds, then unwind the legs and arms and stand in Tadasana again. Repeat for the same length of time with the arms and legs reversed.

Gomukhasana (Cow Face Pose)

5 Simple Steps

1. Sit in Dandasana (Staff Pose), then bend your knees and put your feet on the floor. Slide your left foot under the right knee to the outside of the right hip. Then cross your right leg over the left, stacking the right knee on top of the left, and bring the right foot to the outside of the left hip. Try to bring the heels equidistant from the hips: with the right leg on top, you will have to tug the right heel in closer to the left hip. Sit evenly on the sitting bones.

2. Inhale and stretch your right arm straight out to the right, parallel to the floor. Rotate your arm inwardly; the thumb will turn first toward the floor, and then point toward the wall behind you, with the palm facing the ceiling. This movement will roll your right shoulder slightly up and forward, and round your upper back. With a full exhalation, sweep the arm behind your torso and tuck the forearm in the hollow of your lower back, parallel to your waist, with the right elbow against the right side of your torso.

Roll the shoulder back and down, then work the forearm up your back until it is parallel to your spine. The back of your hand will be between your shoulder blades. See that your right elbow does not slip away from the right side of your torso.

3. Now inhale and stretch your left arm straightforward, pointing toward the opposite wall, parallel to the floor. Turn the palm up and, with another inhalation, stretch the arm straight up toward the ceiling, palm turned back. Lift actively through your left arm, then with an exhalation, bend the elbow and reach down for the right hand. If possible, hook the right and left fingers.

4. Lift the left elbow toward the ceiling and, from the back armpit, descend the right elbow toward the floor. Firm your shoulder blades against your back ribs and lift your chest. Try to keep the left arm right beside the left side of your head.

5. Stay in this pose about 1 minute. Release the arms, uncross the legs, and repeat with the arms and legs reversed for the same length of time. Remember that whichever leg is on top, the same-side arm is lower.

Try to practice these poses every daily, holding the pose 30 to 60 seconds. If the areas you are in a little tight do not worry, you can practice the mountain pose or the eagle pose because they take the least amount of space. Remember, you need to take care of yourself because no one is going to do it for you and stress is the main cause most illnesses and asthma attacks. Stress wears your body and leaves your body open for an asthma attack! Do those poses and relieve that stress.

Chapter 21: Mediation & Asthma

Mediation put you in a total state of relaxation. Mediation can help you prevent unnecessary asthma attacks. When you mediate, you do not have to be focused to achieve the benefits of mediation. You can rest the mind naturally, through total awareness. Some people who practice mediation may ask, "What is she talking about? How am I supposed to rest my mind without focusing on a serene or relaxing image in my mind?"

You can rest your mind, by sitting in a bath, letting out a sigh, and just relaxing in an area in your home that a positive, relaxed, calm energy. This time is called your alone time. There is no standing on your feet, no work clothes on your body, just something that is comfortable. At this point, you need to say goodbye to all your worries. Do not think about work or any responsibilities you need to tackle. Most of all no TV or the computer!

This is how you rest yourself in objectless mediation. You need to just let go and relax. You do not have to stop of block whatever thoughts or emotions that enter your mind, but you do not have to focus on them or follow them. Let yourself rest, by simply allowing whatever to occur. If thoughts and emotions come, let them, be aware of them just do not follow or focus on them. This is called, "Objectless Shinay Mediation."

Objectless Shinay Mediation does not mean letting you mind wonder without direction or putting your mind in fantasies. There is still some presence. This is called center of awareness. You are not focused on anything in particular, but you are still aware. You are aware of what is happening around you.

When you do this, you are resting the mind in its natural simplicity, different from focusing on thoughts and emotions. This is kind of like accepting whatever comes your way, such as accepting whatever ocean wave comes to you. Recognizing that the ocean remains unchanged even though some waves are large and some are small. The ocean waters are always blue and always in motion.

In the same way, this mediation is always clear even when thoughts and emotions pop-up. All the simplicity, clearness, compassion, and calmness are enclosed within that state.

Chapter 22: Helpful Techniques to Deal with Your Stress

In today's world, Stress is unavoidable. Many people suffer from stress. Suffering from stress is very common. Between work, family, relationships, and friendships, how can you avoid stress in your everyday life?

When individuals with asthma experience stress, they increase their chances of having an asthma attack tremendously.

The key is to learn how to deal with your stress and not let it get the best of you. One big mistake that we all make is that we neglect ourselves because we have become so preoccupied focusing on everyone else that we forget about ourselves. The first step to eliminating the stress in your life is learning to take time out for you.

Here are some tips to reduce the stress in your everyday life:

1. Take a hot bubble bath for fifteen minutes. Also, place an oatmeal bath in the water.

2. Lie in the bath tub and close your eyes, take four deep breaths slowly.

3. While you are taking these deep breaths clear all thoughts from your mind. Focus on the feeling of the warm water touching your body and the breathing techniques that you are doing at that moment.

4. Think about something positive and pleasant. Envision something that makes you happy. Focus on something that makes you feel good about yourself.

5. Let go of any negative thoughts that you have stored in your mind. Just to think about one thing that makes you feel good about yourself.

6. Take four more deep breaths relax for a minute and get out of the bathtub.

7. Get dressed, go to a quiet place and sit on the floor. Close your eyes and slowly bend forward, relaxing any tight muscles that are causing you tension. Bend to the left, stretching your arms as far as they will go, then stretch to the right, repeating the movement.

8. Take five more deep breaths and say aloud "I have the patience to change myself and become the person I want to become in life." Say, "I am not going to let the stress I feel control me."

9. Repeat step seven and eight

10. Take five more deep breaths and listen to yourself when you are doing this exercise. Concentrate on yourself while doing this exercise. Do not let any distractions impose on your quiet time. Do not think about anything except this exercise and the techniques it involves.

STEP TWO:

Create a journal in the beginning of the book called the **priority calendar**. Ask yourself these questions.

- What do you regret not having made more time for?

- If you had more time, what would you do with it?

- What are the top ten priorities in your life right now?

- What are your family-related goals?

- What are your business goals?

In the back of the journal, take a quarter of the notebook and title it your **Daily Diary**. Dedicate the diary to you. Write how you felt each day and what made you feel this way. Expressing your emotions on paper helps release your stress and helps heal any unsettled emotions that you may be carrying inside yourself. The goal is to let yourself open up and write intimately and honestly about how you feel.

Working on yourself can be tough if you have a busy schedule; nevertheless, do not let that stop you. You have to make time for yourself. Remember; you come first in life. You need to believe that you are the best. You cannot take care of the people who mean the most to you nor do the things in life that you want to do, if you are having an asthma attack or experiencing asthma symptoms. You need to make sure you feel your best each day; you can do this by eliminating the unwanted stress in your life.

On the weekends, take time to reward yourself. For example, take in a movie or reward yourself with some quiet time to relax and focus just on yourself. To me there is nothing better than having some time alone (quiet time). Do something that makes you happy. Remember, you cannot make others happy until you are happy with yourself. You are not going to be happy if you experience an asthma attack.

Develop a special time in the day for quiet time. Studies have shown that individuals who have a daily quiet time are less likely to become have an asthma attack. Take a few minutes during the day to write in your journal. Try to make it the same time each day. You could do this when no one is home or just before you start the day in the morning. You could also wait until everyone goes to sleep so that no one will bother you. Give yourself at least fifteen minutes to a half hour. Think of ways to strengthen yourself spiritually and emotionally. Make sure you do not limit yourself because you feel sorry for yourself because you have a stressful life or because you have asthma. That is self-pity and it is unhealthy. You will never get anywhere in life if you pity yourself. Free yourself from any walls you have built around yourself.

Now repeat this process each day and review the things you have written in your journal. These techniques will help you reduce the stress in your life. Keep doing these techniques each day until you feel better and you have become completely satisfied with yourself. You should always work on bettering yourself. Everyone is special and needs to take time out for him or herself. There is no excuse for neglecting yourself.

Section Five - Resources: Asthma Support
Groups & Organizations

Chapter 23: Support Groups and Asthma Organizations That Can Help You and Your Loved Ones

Resources on the Worldwide Internet for Asthma

Allergy and Asthma Network/Mothers of Asthmatics America, Inc. (AAN/MA)
8201 Greensboro Drive, Suite 300
McLean, VA 22102
(800) 878-4403
www.aanma.org

<u>American Academy of Asthma, Allergy and Immunology (AAAAI)</u>
555 East Wells Street
Suite 1100

Milwaukee, WI, 53202-3823
(414) 272-6071
www.aaaai.org

American Association for Respiratory Care (AARC)
9425 N. MacArthur Blvd. Suite 100
Irving, TX, 75063-4706
(972) 243-2272
www.aarc.org
E-mail: info@aarc.org

Asthma and Allergy Foundation of America (AAFA)
8201 Corporate Drive
Suite 1000
Landover, MD 20785
1-800-7-ASTHMA (727-8462)
E-mail: info@aafa.org
www.aafa.org

American College of Chest Physicians
3300 Dundce Rd.
Northbrook, IL 60062-2348
(847) 498-1400 or 1-800-343-2227
www.chestnet.org

American Lung Association
1301 Pennsylvania Ave. NW
Suite 800
Washington, DC 20004
(202) 785-3355
E-mail: info@lung.org
www.lung.org

The Food Allergy & Anaphylaxis Network
11781 Lee Jackson Hwy., Suite 160

Fairfax, VA 22033-3309
(800) 929-4040
www.foodallergy.org

Natural Heart, Lung and Blood Institute
This is the division of the US National Institutes of Health (NIH) that deals with respiratory illnesses such as asthma. Provides easy-to-understand information on asthma and how to manage it. They also publish research studies. I find the NIH to be the most reliable source of health information on the Web. They also do a great job of making their information understandable for the layperson.

The Lung Association of Canada
The Lung Association is an organization with a mission of promoting and improving respiratory health in Canada. You can find asthma information, news headlines, and tips on how to get help for asthma in Canada.

Centers for Disease Control and Prevention
This website provides general information and information on asthma treatments. I recommend visiting the site if you are interested in getting the latest information from the US Government on environmental health concerns related to asthma. There are also even more links to some great asthma resources.

National Jewish Medical Center
This is another great resource for learning about asthma and how to manage it. One great feature is the "Asthma Wizard", which is an animated guide to learning about asthma that is suitable even for children.

American Lung Association

American Lung Association is the oldest voluntary health organization in the United States. They provide information on asthma, allergies, and other respiratory illnesses as well as ways to stop smoking. For in-depth information on ongoing clinical trials, go to the American Lung Association's Clinical Research Centers.

American Academy of Allergy Asthma & Immunology

American Academy of Allergy Asthma & Immunology is one of the nation's largest professional medical specialty organizations. Their journal is highly respected worldwide, with many studies about asthma each month.
Asthma and Allergy Foundation of America

National Heart, Lung, and Blood Institute

The National Heart Lung and Blood Institute plans, conducts, and supports research, clinical trials, and observational studies related to the causes, diagnosis, prevention, and treatment of heart, blood vessel, lung, and blood diseases, and sleep disorders.

The National Institute of Allergy and Infectious Diseases

The National Institute of Allergy and Infectious Diseases conducts research in allergic asthma and infections, such influenza. The NAIAD funded the National Inner City Asthma Study, which determined that control of indoor allergens improved asthma control in children living in cities across the U.S.

The Environmental Protection Agency (EPA)

This agency has funded many studies about the dangers of outdoor air pollution, such as ozone, NOX, sulphur dioxide, and smoke, for those with asthma. They produce "Tools for Schools," a great guide for reducing asthma triggers in schools, and have a fun Kids Club for sleuthing causes of environmental disease.

Asthma Foundation of Victoria

Website address: www.asthma.org.au

Asthma UK

Nonprofit U.K. organization working to conquer asthma through a combination of research, education, and support to those affected by asthma. Site features general information, trigger tips, latest research news, discussion forum, advice line, as well as online resources for patients and professionals.

Website address: www.asthma.org.uk

National Asthma Educator Certification Board

Promotes optimal asthma management and quality of life among individuals with asthma through the Certified Asthma Educator process.

Website address: www.naecb.org

Asthma Society of Ireland

This support organization provides information, advice, and reassurance to people with asthma and their families.

Website address: www.asthmasociety.ie

Global Initiative for Asthma

Provides access to latest World Health Organization and NIH guidelines for asthma management for health professionals and patients.
Website address: www.ginasthma.com

Asthma Society of Canada
Presents a guide on how to live with asthma through education and proper treatment.

Website address: www.asthma.ca

Asthma Information Outreach
Includes statistics and information about programs, projects, and events in the New York City area.
Website address: www.asthma-nyc.org

European Federation of Asthma and Allergy Associations
This site promotes better health for people with asthma and allergy throughout Europe.

Website address: www.efanet.org

National Asthma Campaign (NAC) - Australia
Provides information for health professionals and people with asthma and their caregivers.

Website address: www.nationalasthma.org.au

Asthma and Allergy Foundation of America- Maryland Chapter
Dedicated to improving quality of life for asthma and allergy sufferers.
Website address: www.aafa-md.org

"There is no cure for asthma. Your comfort and peace of mind depend on how well you learn to live with it."

About the Author

Stacey Chillemi

Stacey Chillemi graduated from Richard Stockton College in Pomona, New Jersey, majoring in marketing and advertisement. In the mid-nineties while in college, she began her first book, *Epilepsy: You're Not Alone.* It was published six years later. Before and after graduation in 1996, she worked in New York City for NBC. Since the birth of her children, she has been a freelance journalist.

She has written features for journals and newspapers. Her articles have appeared in dozens of

143

newspapers and magazines in North America and abroad. She won an award from the Epilepsy Foundation of America in 2002 for her help and dedication to people with epilepsy.

My Web Sites:
http://www.lulu.com/spotlight/staceychil
http://www.authorsden.com/staceydchillemi

BOOKS PUBLISHED BY STACEY CHILLEMI:

- The Complete Herbal Guide: A Natural Approach to Healing the Body
- How to Live Comfortably with Asthma
- Epilepsy You're Not Alone
- Eternal Love: Romantic Poetry Straight from the Heart
- My Mommy Has Epilepsy (Children's Book)
- My Daddy Has Epilepsy (Children's Book)
- Keep the Faith: To Live and Be Heard from the Heavens Above (poetry book)
- Live, Learn, and Be Happy with Epilepsy
- Epilepsy and Pregnancy: What Every Woman Should Know
  Co-authored by Dr. Blanca Vasques
- Faith, Courage, Wisdom, Strength and Hope
- How to Be Wealthy Selling Informational Products on the Internet
- How to Become Wealthy in Real Estate
- How to Become Wealthy Selling Ebooks

- Life's Missing Instruction Manual: Beyond Words
- How To Become Wealthy Selling Products on The Internet
- Breast Cancer: Questions, Answers & Self-Help Techniques
- How Thinking Positive Can Make You Successful: Master The Power Of Positive Thinking
- Beginners Tips for Horse Training: What Every Horse Trainer Should Know
- Natural Cures for Common Conditions: Learn How to Stay Healthy and Help the Body Using Alternative Medicine, Herbals, Vitamins, Fruits and Vegetables
- The Ultimate Guide to Living Longer and Feeling Younger
- How to Buy a Home Using a VA Loan: What Every Home Buyer Should Know

STACEY CHILLEMI STORIES AND POETRY HAVE BEEN PUBLISHED IN:

- **Chicken Soup for the Recovering Soul**
- **Chicken Soup for the Shoppers Soul**
- **Whispers of Inspiration**

ACCOMPLISHMENTS:

- Book Signing at Borders in Freehold, New Jersey for Faith, Courage, Wisdom, Strength and Hope" – July 2009

- Writer for Neurology Now Magazine (The Academic Academy of Neurology – The Epilepsy Column) February 2010
- February 2010, Wrote an article about Epilepsy & Menstruation with Dr. Devinsky (Epileptologist from NYU)
- H.O.P.E. Mentor, for the Epilepsy Foundation
- Speaker at different events for schools, organizations, political events
- Spoke in front of Congress in Washington for employment discrimination for people with epilepsy
- Appeared on four talk shows to discuss epilepsy focusing on the importance of understanding epilepsy, how to help someone having a seizure and giving people with epilepsy encouragement and hope for the future.
- Appeared on radio stations discussing epilepsy
- Appeared on the Michael Dressor Show – Health Radio
- Appeared in newspapers all over New Jersey such as, The Leader, Belleville Post and the Star Ledger.
- June 26, 2002, I was honored an award by the Epilepsy Foundation of New Jersey for Outstanding Volunteer Award.
- Received awards in my achievements and certificates in recognition for outstanding efforts in trying to improve society.
- Active participant in organizations and activities.
- Published over 400 articles.

CAREER EXPERIENCE:

- Journalist for The Journal Magazine
- Worked for NBC on Dateline
- Channel 4 News
- Today Show
- Managing Editor for the Fashion Magazine **UZURI**.
- Own Freelance Company

References

Smolley, L. and Bruce, D. Breathe Right Now. (New York: Dell, 1998). Bruce, D. The Sinus Cure. (New York: Ballantine, 2007). American Academy of Allergy, Asthma & Immunology: "Tips to Remember: asthma triggers and management." American Academy of Allergy, Asthma & Immunology: "Allergic Conditions: Exercise-Induced Asthma (EIA)." Merck Manual Home Edition: "Asthma." American Academy of Allergy Asthma & Immunology (AAAAI): "An Unwelcome Return: 10 tips to ease your spring allergy symptoms."

:American Academy of Allergy, Asthma and Immunology web site: "GERD and Asthma," "Sulfite Sensitivity," "Tips to Remember: Asthma Triggers and Management." American Medical Association, Essential Guide to Asthma, 2000. News release, American Thoracic Society web site. Freary, J. and Britton, J. Thorax, 2007; vol 62: pp 466-468. McKeever, T.M. and Britton, J. American journal of Respiratory and Critical Care Medicine, 2004; vol 170: pp 725-29. Medscape Medical News: "Mediterranean Diet May Protect Against Childhood Asthma-Like Symptoms and Rhinitis." National Institute of Allergy and Infectious Diseases web site: "Food Allergy: An Overview." National Jewish Medical and Research Center: "Nutrition and Asthma." WebMD Medical News: " Poor Diet May Affect Teen Asthma."

The Lung Association: "Exercise & Asthma." American Academy of Allergy, Asthma & Immunology: "Tips to Remember: Exercise-induced asthma." Asthma and Allergy Foundation of America: "Exercise-Induced Asthma."

American Academy of Family Physicians: "Stress: How to Cope Better with Life's Challenges." The National Women's Health Information Center: "Stress and Your Health." Barclay, L. Thorax, June 2007.
Photo by adam. Normal and abnormal bronchiole

Photo nubulizer by adam

Photo peak flow meter adam

Elana Pearl Ben-Joseph, MD : May 2007

MedicineNet.com: "Definition of Beta-Agonist." National Jewish Medical and Research Center: "Inhaled Medication with a Metered Dose Inhaler (MDI.)" Asthma Society of Canada: "How to Use Your Inhaler." Science Daily: "New Asthma Inhaler Propellant Effective, but Costlier." Children's Hospital Boston: "Allergy Treatment."

Asthma UK: 23 October 2007

Images WebMD 2012
Melinda Ratini, DO, MS on July 10, 2012

ISBN: 978-1-300-40745-4

www.ingramcontent.com/pod-product-compliance
Lightning Source LLC
Chambersburg PA
CBHW021951170526
45157CB00003B/937